DINOSAUR
Safari
AF258901
2021
Coloring
Calendar

Copyright 2020 Gumdrop Press

All rights reserved.

No part of this book may be reproduced in any written, electronic, or photocopied form without written permission of the publisher or author.

Every effort has been made to ensure the accuracy of the information contained in this book. The author and publisher disclaim liability to any party for any loss, damage, or disruption caused by errors or omissions that may result from the use of information contained within, whether such errors or omissions result from negligence, accident, or any other cause.

JANUARY 2021

Sun.	Mon.	Tue.	Wed.	Thu.	Fri.	Sat.
					1 New Year's Day	2
3	4	5	6	7	8	9
10	11	12	13	14	15	16
17	18 Martin Luther King, Jr. Day	19	20	21	22	23
24	25	26	27	28	29	30
31						

FEBRUARY 2021

Sun.	Mon.	Tue.	Wed.	Thu.	Fri.	Sat.
	1	2	3	4	5	6
7	8	9	10	11	12	13
14 Valentine's Day	15 Presidents' Day	16	17	18	19	20
21	22	23	24	25	26	27
28						

MARCH 2021

Sun.	Mon.	Tue.	Wed.	Thu.	Fri.	Sat.
	1	2	3	4	5	6
7	8	9	10	11	12	13
14 Daylight Saving Time Begins	15	16	17 St. Patrick's Day	18	19	20
21	22	23	24	25	26	27
28	29	30	31			

APRIL 2021

Sun.	Mon.	Tue.	Wed.	Thu.	Fri.	Sat.
				1	2	3
4 Easter	5	6	7	8	9	10
11	12	13	14	15	16	17
18	19	20	21	22	23	24
25	26	27	28	29	30	

MAY 2021

Sun.	Mon.	Tue.	Wed.	Thu.	Fri.	Sat.
						1
2	3	4	5	6	7	8
9 Mother's Day	10	11	12	13	14	15
16	17	18	19	20	21	22
23	24	25	26	27	28	29
30	31 Memorial Day					

JUNE 2021

Sun.	Mon.	Tue.	Wed.	Thu.	Fri.	Sat.
		1	2	3	4	5
6	7	8	9	10	11	12
13	14	15	16	17	18	19
20 Father's Day	21	22	23	24	25	26
27	28	29	30			

JULY 2021

Sun.	Mon.	Tue.	Wed.	Thu.	Fri.	Sat.
				1	2	3
4 Independence Day	5 Independence Day Observed	6	7	8	9	10
11	12	13	14	15	16	17
18	19	20	21	22	23	24
25	26	27	28	29	30	31

AUGUST 2021

Sun.	Mon.	Tue.	Wed.	Thu.	Fri.	Sat.
1	2	3	4	5	6	7
8	9	10	11	12	13	14
15	16	17	18	19	20	21
22	23	24	25	26	27	28
29	30	31				

SEPTEMBER 2021

Sun.	Mon.	Tue.	Wed.	Thu.	Fri.	Sat.
			1	2	3	4
5	6 Labor Day	7	8	9	10	11
12	13	14	15	16	17	18
19	20	21	22	23	24	25
26	27	28	29	30		

Dinos
Pteradactyls
T-Rex
Humans

OCTOBER 2021

Sun.	Mon.	Tue.	Wed.	Thu.	Fri.	Sat.
					1	2
3	4	5	6	7	8	9
10	11 Columbus Day	12	13	14	15	16
17	18	19	20	21	22	23
24	25	26	27	28	29	30
31 Halloween						

NOVEMBER 2021

Sun.	Mon.	Tue.	Wed.	Thu.	Fri.	Sat.
	1	2 Election Day	3	4	5	6
7 Daylight Saving Time Ends	8	9	10	11 Veterans Day	12	13
14	15	16	17	18	19	20
21	22	23	24	25 Thanksgiving Day	26	27
28	29	30				

DECEMBER 2021

Sun.	Mon.	Tue.	Wed.	Thu.	Fri.	Sat.
			1	2	3	4
5	6	7	8	9	10	11
12	13	14	15	16	17	18
19	20	21	22	23	24 Christmas Observed	25 Christmas
26	27	28	29	30	31 New Year's Day Observed	

www.ingramcontent.com/pod-product-compliance
Lightning Source LLC
Chambersburg PA
CBHW080312030726
47593CB00009B/2732